Making Your Camera Pay

Frederick C. Davis

Alpha Editions

This edition published in 2022

ISBN : 9789356715981

Design and Setting By
Alpha Editions
www.alphaedis.com
Email - info@alphaedis.com

Contents

WHAT IT'S ALL ABOUT

Whence come the thousands of photographs used every month by newspapers and magazines?

More than that, whence do the photographs come which are used by makers of calendars, postcards, for advertisements, and for illustrating books, stories and articles?

At first thought, the answer is, "From professional photographers and publisher-photo-services." But professional photographers do not produce one-third of the photographs used, and publisher-photo-services are supplied by that same large number of camerists that supply publications with most of their prints.

No one can deny that the greatest number of prints published are bought from amateur photographers in towns no larger than the average, and sometimes smaller.

The camerist does not have to get in an air-ship and fly to Africa in order to produce photographs that will sell. Read what Waldon Fawcett says, himself a success at selling his photographs:

"The photographer is apt to think that all his ambitions would be realised if only he could journey to foreign shores or to distant corners of our country; or if he could attend the spectacular events that focus the attention of the world now and then. *This is a delusion. The real triumph is that of the photographer who utilises the material ready at hand in his own district, be it large or small.*"

And more, a person does not have to be an expert photographer in order to succeed at the work. Here is what one prominent writer says about it:

"The requirements of the field are well within the capabilities of even the beginner in photography, viz.; the ability to make good negatives and good prints, the ability to recognise news-value, and a methodical plan to find the market where the prints will find acceptance. The man or woman who can meet these requirements should be fairly successful from the beginning, and will open up quickly new avenues of special work and profit."

In short, ability to make metaphors, create lovely heroines or such is not at all necessary to the successful selling of photographs to publications.

Is the field overcrowded? *No.* If there were ten times as many persons engaged in the work they could all keep themselves busy.

The field—how wide is it? Get out your map of the world. The field for *making* photographs extends from the top margin to the bottom, and from the left to the right. The field for *selling* photographs—which is more to the point—extends over about five thousand publications which use prints; not to speak of a few score of other markets.

The markets may be classified briefly:

- (1) Newspapers
- (2) Magazines
- (3) Postcard-makers
- (4) Calendar-makers
- (5) Art-study producers
- (6) Illustrations for books
- (7) Illustrations for articles
- (8) Prints for advertising.

And there are more, of more specialised branches.

And how does it pay? Please note: "A certain magazine once paid $100 for four prints of sundials. An amateur, who happened to be on the spot with a kodak, made over $200 out of a head-on railroad-collision. A New York professional netted $125 from the newspaper-use of a wedding-party, of considerable local prominence, which was leaving the church after the ceremony." One amateur "realised $300 a year for two or three years from a lucky snapshot of eight pet rabbits in a row."

A set of South-Pole photographs brought $3,000 from *Leslie's* and $1,000 more from the International Feature Service. These all, though, are very exceptional instances. The average print sells for about three dollars. But there is absolutely nothing in the world to hinder a wide-awake person with a camera from making from several hundred to over $3,000 a year from his prints. If he becomes a specialist he may earn as high as $5,000 or even more.

No discrimination is made between press-photographers. The person wins who "delivers the goods."

However, I do not mean that the instances of $200 or so for prints should be taken as the prices ordinarily paid. I do not maintain that there is a fortune awaiting the man with the camera; but I do say there are unlimited possibilities for salable photographs and almost an unlimited number of markets for them. But there are not "barrels of money" in it, for all. A person may add appreciably to his income for having sold photographs; and having developed the trade to a high degree, he may cash cheques to the amount of

$5,000 or more a year. But not every one. Just some. And it isn't like the log and the falling off it. It's work—hard work—*hard work*.

Success at selling press-photographs does not depend on the size of the town you live in, the cost or manufacture of your apparatus, or on your literary ability. It depends on you and your worship of the homaged gods of success if you would sell photographs. The gift of these gods is the ability to make good.

THE TOOLS OF THE TRADE

Have you ever wakened in the drear dead of a dismal night, possessed body and soul with a great desire—an incontrollable, all-moving, all-consuming, maddening desire that knows no satisfaction—a desire for a new camera or a better lens? It is a sensation more disconcerting than that of the father who is detected by his small son in the act of rifling the latter's bank for car-fare. Never would I be so unwise as to cultivate that desire in any one; for that reason I do not here go deeply into a discussion of the best kind of camera for press-photography! Unless the camera you now possess is of a hopelessly mediocre grade, it will do very well.

A reflex camera is of course the ideal instrument for the purpose, for sharp focusing is so easy and so necessary. The high speeds of the focal-plane shutter incorporated into such a camera will rarely be utilised by the average user; but its other features are admirable.

However, the hand-camera of the folding type is supreme. It is so light it can be carried for a long time without fatigue; the user of one is inconspicuous when making exposures; the cost of operation as well as the original outlay is comparatively small—and there are several dozen more things in favor of it, including its greater depth-of-field, which is most important.

The lens is the heart of the camera, and some cameras have "heart-trouble." If you intend seriously to market photographs you should possess an anastigmat lens; not necessarily an F/4.5 lens, nor even an F/6.3 lens if too expensive; in that case an F/7.5 lens will do very well. An F/7.5 anastigmat is slightly slower than a rapid-rectilinear of U.S.4 aperture; but its excellence lies in its ability—as with all anastigmats—to form images of razor-edge sharpness, which is a prime requisite of a print intended to grace a page of a periodical. A rapid-rectilinear lens will do very well if you are always assured of sunshine or bright clouds to supply exposure-light—and in such conditions even the lowly single-achromatic lens will suffice.

Now you see I have agreed that virtually any lens that will form a sharp image will meet the requirements. Indeed, to paraphrase Lincoln: "For the sort of thing a lens is intended to do, I would say it is just the lens to do it." In other words, each lens has its limitations and abilities very sharply defined; and these limits the user must know and appreciate.

And the shutter; it is folly to put a poor lens in a good shutter, and just as absurd to do the opposite. An expensive shutter with high speeds cannot be successfully used except with a lens capable of large aperture—otherwise underexposure will result. A speed of 1/300 second is the highest available in an ordinary between-the-lens shutter, and that is sufficient for almost anything.

The slower speeds, as one-fifth, one-half and one second are in my opinion more usable than the extremely fast ones. Speeds varying from one second to 1/300 second are embodied in two well-known shutters: the Optimo and the Ilex Acme. The one is on a par with the other. But no such high-grade shutter is needed unless the high speeds are necessary to the user, for the slower speeds may be given with the indicator at B. But enough! This is not a manual on the elements of photography.

The requirements of the apparatus to be used for press-photography are that the lens produce a sharp and clear image, the shutter work accurately, and the whole be brought into play quickly.

I have used every sort of camera; reflex, 8 × 10 view, 5 × 7 view, hand-cameras with anastigmat, rapid-rectilinear and single lenses, and box-cameras, and they are all entirely satisfactory "for the things they were intended to do."

The camera I have used most and which is my favorite is a Folding Kodak, that makes 3¼ by 4¼ photographs, and is equipped with an Ilex Anastigmat working at F/6.3, in an Ilex Acme shutter. To this I have added a direct-view finder for reasons apparent to any one who has tried to photograph high-speed subjects by peeking into the little reflecting-finder. This camera has served me admirably for interiors, flashlights, outdoors, high-speed work, portraiture, and anything else to which I have applied it. Your own camera should do the same for you.

A photographer comes to know his camera as a mother knows her baby—and if he doesn't he will be no more successful than the mother who does not understand her child. The camera-worker must forget all about manufacturers' claims and should judge his tool by experience; he must ignore most of the theory and rely wholly on practice. In short, he must know his camera inside and out, what it will do and what it will not do; everything must be at his finger-tips ready for instant use. Coupled with that is the need of the ability to produce, sometimes, within an hour after making the exposure, crisp, sharp, sparkling prints.

After all, no more qualifications are required of the press-photographer than of most other photographers. He may have to work like lightning, snap his shutter literally under the very hoofs of racing-horses, rush out of a warm and cozy bed into a chill and bleak night—but "it's all in the game." If any one of the old veteran press-photographers were to lead the life of an ordinary business-man, he would die of ennui. When the camerist makes photographs for publishers it is zip-dash—and later, cash.

It is the exciting life of a never-sleep reporter, with a camera to manage instead of a pencil.

WHAT TO PHOTOGRAPH

If you wish immediate wealth you have only to locate several oil-pockets and dig into them. Similarly, if you aspire to success at marketing photographs you have only to discover the needs of editors and to satisfy them. But although there are not many more available oil-pockets, there are many editors and innumerable editorial needs.

It would be as absurd for me to attempt to state precisely what you should photograph as it would be for me to make a pencil-dot on a map and to say: "There's an oil-pocket; go dig into it." The one way to discover the needs of editors and how to satisfy them is to develop a "nose for news."

A "nose for news" is simply the ability to determine the value of any certain photograph to any certain editor. The several ways of acquiring that very necessary ability are: (*a*) by experience, which consumes the most time and is the most difficult; (*b*) by examining the nature of photographs already sold to publications and printed in them, which is less difficult and just as effective; and (*c*) by careful study of prevailing editorial needs and market-demands, which is the best method of all.

To succeed, mix thoroughly liberal quantities of (*a*), (*b*) and (*c*).

Not many, other than the large metropolitan newspapers, employ staff-photographers; and if a smaller one does, the photographer is usually a reporter who has much scribbling to do besides. When most newspapers require a photograph of something local, the city-editor telephones to a commercial-photographer and tells him to "get it." Thereupon, the commercial-photographer packs up his forty-pound outfit, goes out and gets it.

However, a good many subjects are not of sufficient interest to cause the city-editor to dispatch a commercial-photographer to obtain them; but, if photographs of those same subjects were brought unsolicited to him he would at once see their value and buy them. That is the biggest advantage of the free-lance photographer with the newspapers.

If the press-photographer wishes to follow these tactics he may profit, even in a very large city; for staff-photographers go where city-editors tell them to go, and city-editors have much to think about.

The kinds of subjects bought by newspapers from free-lance photographers are those of local interest, brought to the office while the interest in them is still keen. A large number of such subjects are available daily. The news-photographer may glean his tips from a morning-newspaper and sell his prints to an evening-journal. When he becomes sufficiently well known, he may be called upon and dispatched after a photograph just as the commercial-photographer. But first he must impress the editorial mind by giving it, unasked, the very sort of thing it wants.

The free-lance photographer should see possibilities in many subjects:

- A public building burns.
- A corner-stone is laid.
- An illicit still is found.
- A new building is erected.
- A murder occurs.
- A new fire-department truck is bought.
- The governor comes to town.
- Josh Jones finds a hen's egg three-times normal size.
- A park is improved.
- The first baseball-game is played.
- The robber of the postoffice is caught.
- I. Wright, the local author's new book, is published.
- The local inventor again invents.

Any one of these suggestions holds possibilities for photographs useful to a newspaper; and many more events are just as promising.

The types of photographs used by postcard-makers are known to almost every one. The subjects run from famous buildings and historical monuments to artistic human-interest pictures such as a small kitten sleeping with its feet entangled in a maze of thread with which it has been playing.

At that point, merge the demands of the calendar-makers. They use the human-interest type, and run to landscapes, seascapes, and portraits of pretty girls. Usually the demand of both postcard- and calendar-makers is that the picture tell a story. If it can be used without an explanatory caption, all the better. For an example of a picture-told story, glance at almost any cover of the *Saturday Evening Post* and note how the whole situation is made clear without one word of explanation. It is that kind of photograph that postcard- and calendar-makers want. If you will glance over the postcard- and calendar-illustrations you have at hand you will readily see the types of photograph used.

Sometimes book-publishers send out calls for special kinds of photographs they need in preparing certain books. In that case, they usually advertise in an appropriate magazine and mention the kind of photograph they wish; for example, historical prints if a history is in preparation. The unlimited variety of books published calls for an unlimited variety of photographs. Certain publisher-photo-services make it their business to supply publishers with the photographs they wish; but that is not hurtful to the prospects of the free-lance, for the photo-services must obtain photographs of every kind from every source, and must be stocked with a larger number and variety of prints than any one magazine or publisher could possibly use. Thus, in fact, the news-photographer has an increased market.

The largest field for the free-lance photographer I have left until last; that is, the magazines. There are so many magazines and such a variety of them that almost any print, if it is of interest at all, should find a place with one of them. Besides the large magazines there are many smaller ones; those devoted to almost any conceivable vocation, and others to almost any interest or hobby.

Besides the publications issued for the great mass of the reading public, there are magazines published solely for advertisers, architects, real-estate agents, automobilists, bakers, confectioners, cement-users, drug-stores, dry-goods merchants, electricians, engineers, miners, bankers, financiers, fraternal members, furniture-dealers, millers, grocers, hardware-sellers, historians, hotel-owners, owners of restaurants, jewelers, labor-union members, lawyers, insurance-agents, soldiers, sailors, municipal workers, printers, publishers, railroad men, magicians, fox-raisers, blacksmiths, fruit-growers, undertakers, stamp-collectors, and scores of others, not to speak of almost two thousand house-organs issued by manufacturers as sales-promotion literature or for the benefit of their employees. And each of these uses photographs occasionally, if not regularly. The photographer need not deplore a lack of sufficient markets for his photographs.

The greatest influence toward the development of a "nose for news" is the giving to it of several whiffs of news. A photographer may "shoot"—a professional photographer never photographs—he shoots—he may shoot and shoot, and have his every photograph returned to him as useless for publication—but not if he first discovers what to photograph and what not to photograph.

As a means toward that end I have selected, at random, issues of three magazines whose pictorial sections contain prints which are, broadly, just the sort of photographs the photographer in a medium-size town produces. The magazines are *Popular Science*, *Illustrated World*, and *Popular Mechanics*; despite their names, these magazines print photographs of a very general scope—more general than one would suppose. I have selected only photographs with short captions, or those with explanatory articles not more than two hundred or so words in length.

In *Popular Science* I find:

- An Apartment-House for Plants.
- A Hospital on Wheels.
- Potato-Gathering Made Easy.
- This Rudder Makes the Boat Behave.
- New Light for the Photographer.
- He Wears a Showcase.
- A Rubber Heel with a Noise.
- Milking Cows by Electricity.
- Anchoring Bricks to the Side of a House.
- Sketching on Fungus, One Artist's Hobby.
- Sampling the Soil.
- Making House-Wrecking Easy.
- A Machine that Harvests Crimson Clover Seed.
- Wheel-Guards that Save Life.
- Working Safely on High Voltage Lines.
- A Lake that has a Crust of Salt.
- Punching Your Votes.
- Your Money is Safe in this Bank-Tank.

In *Illustrated World*:

- Motorized Wheel-Chair for Invalids.
- Whirr of Motors Replaces Song of Cotton-Pickers.
- How Aristocrats of Dogdom Travel.
- Perform Marriage-Ceremony in Oil-Filling Station.
- Rail Motor-Trucks for Short-Line Road's Use.
- No More Backaches from the Lawn-Mower.
- Novel Arrangement of Air-Hose for Work-Benches.

- Largest Milk-Tank in the World.

- Comfortable Footrest for a Rustic Seat.

- Dog Hurt in Auto Accident Wears Wooden Leg.

- Street-Cars Adopt "Pay-As-You-Leave" System.

- Dentists' Scales for Weighing Mercury.

- Toy Makes Spelling Easy for Kiddies.

- Small Check-Book in Silver-Case.

- Nine-Story Building Collapses.

- Traveling Mail-Box on Interurban Car.

- Clever Method of Advertising Perfume.

- Makes Suit Out of Stamps.

- Wellesley Girls Have a "Sneezing Closet."

- Raising Chickens on a Back Porch.

In *Popular Mechanics*:

- Owner of Artificial Hands is Proud of Dexterity.

- Imperishable Burial Robes Shown on Living Models.

- Novelty Window-Sign Spells Words with Snowflakes.

- Imposing New Bridge at Jacksonville.

- Street-Sign Calls for Help if Robbers Invade Store.

- New Style Log-Cabin Built Like Stockade.

- Vines Completely Cover Office-Building.

- Beautiful Ice Stalagmites are Pranks of Jack Frost.

- Unique Wood-Sculptures are Work of a Decade.

- Electric Warehouse-Truck Performs Heavy Tasks.

- Hydraulic Jack Tears Up Street-Car Tracks.

- Man-Power Onion-Planter Sets an Acre a Day.

- Grotesque Images Reward Motor-Cycle Race Winners.

- Weak Derrick Starts Work of Steel-Building.

- Concrete Logging Piers are Used in Lumber-Industry.

- World's Largest Clock Keeps Accurate Time.

- Grotesque Face on Auto Advertises Carnival.

- River-Bed Proves to be a Rich Coal-Mine.

- Outlets of Odd Shapes Made for Irrigation.

- Unusual Park-Playground Built in Circus-Form.

- Giant Vase, Lawn-Ornament, is Made of Concrete.

- Old Silo in Railroad-Yard Houses Little Store.

- Street Rises so Abruptly Four Flights of Steps are Necessary.

- Church Uses Bill-Board to "Sell" Scriptures.

This wide variety of subjects cannot but serve to show that even in very small towns there are many opportunities for salable pictures. More than that, there are markets for prints of:

Statues	Farm-scenes
Blacksmith-shops	Mural decorations
Farm light-plants	Seascapes
Sheep	Gardening operations
Landscapes	Interior decorations
Paintings	Designs
Girls' heads	Camping-scenes
Farm-buildings	Trapped wild animals
New inventions	Freaks
New achievements	Cattle
Live game	Orchards
Birds in flight	Time-saving plans
Industrial arts	Social progress
Fields of grain	Fashions

Desert-views	Wharves
Domestic animals	Paint-departments
Poultry	Mills
Harbors	New banks
Garage-methods	Large estates
Railroading	Factory-equipment
Concrete-construction	Show-window displays
Flowers	Store-fronts
Electrical appliances	Motorcycles
Live-stock prize-winners	Economic interest
Art-museums	Good and bad roads
Motorboats	Spraying-methods
Musical work	Counter-displays
Shoe-factories	Blasting
Prize-dogs	Landscape-gardening
Yachts	Sports

If you live in a large city you have the additional opportunities to obtain photographs such as are published in the *Mid-Week Pictorial* and the *Illustrated Review*, and also in some of the large national magazines and in the rotogravure-sections of the leading Sunday newspapers. Although the large city offers more opportunities for photographs of celebrities and such, there is much competition. The photographer in an average-size city may not have frequent opportunities for photographs of renowned persons; but he has many other chances for salable photographs, which evens up things.

Sometimes, a notable person does come to town; but I would no more presume to tell you here to camp on his trail than I would dare to remark to a duck-hunter: "Pardon me, old man, but you'd better pull your trigger. There's a bird right where you've pointed your gun."

WHAT NOT TO PHOTOGRAPH

Knowing *what* to photograph is no more important than knowing *what not* to photograph. I cannot show you so easily by example the kind of photographs editors will not buy; for a search of any number of magazines will fail to unearth such examples.

Experience is an expensive school; but, sometimes, the others are closed because of lack of patronage. It would seem that when you learn *what* to photograph you should learn automatically *what not* to photograph; and, indeed, you should; but you don't. However, there is another way. After sending a photograph to a score of publications, and after the photograph is returned from the same score of publications, you may truthfully say: "Well, I've discovered one thing that those editors don't want."

Editors have very clear reasons why they don't buy certain kinds of photographs. The editor is there to produce a live, newsy, unusual publication. He buys only live, newsy, unusual photographs. What could be simpler?

Publications do not want photographs which are similar to other photographs that they have already printed. The reason is obvious. To take an example from my own early days: a shoe-dealer, for an advertisement, placed a huge pair of shoes, size 35, in his window. I grasped the opportunity to make a salable photograph. It did sell; but not to *Popular Mechanics*, for the editor wrote that he was unable to use it because he had printed, several months before, a picture of a huge pair of shoes made for a circus sideshow worker. Consequently, the subject of your photograph may be just the thing the editor would want if he hadn't had his requirements already satisfied. Therefore, study those photographs which have been printed, and make newer and better ones.

When the King of England comes to town, it may be all very well to command him to stand still, to look serious or to smile, for a picture of him so posed may be literally "eaten up" by the local newspapers; but a national weekly, such as *Collier's*, demands something different. Posed photographs are at a discount. They are too plainly "pictures of men having their pictures made." What is wanted are life and action. It isn't necessary to ask the King to stand on his head. Ask him to shake hands with the Chief-of-Police; or let him do something else which shows he has the power of action.

On an invaluable rejection-slip prepared by a national magazine, examples are given of "What we want and don't want." Under a photograph of Senator Johnson with upraised fist, as if he were driving home a point in his speech, is printed: "Here the upraised fist does the business—makes action, life—

and transforms what would otherwise be just an ordinary likeness of Senator Johnson into a striking and arresting picture."

But if a photograph is sufficiently unusual it may be without life and yet may sell, although it gains materially by a show of action. Under a photograph of a floating submarine, the rejection-slip notes: "No action here; but it is safe to say that few of the readers of this magazine skipped this one when it appeared. Submarines are common today; but not the kind that carry huge twelve-inch guns." Similarly under a photograph of three men standing in a row and looking with a "where's-the-birdie?" expression at the camera, the caption is: "A posed picture and, as is usual in such circumstances, a dead one. We used it because a story centering around these men was a singularly interesting one appealing to a large audience in America." But no matter how extraordinary a photograph is, it gains a hundred-fold by exhibiting signs of *life*.

True, a "dead" picture may sell; but a live one will sell more quickly, and the photographer's work will be more in demand, and the resulting cheque will be larger—much larger.

If you make a photograph of a building—even for instance, a new arsenal—you will never sell it to such a publication as the New York *Times* roto-section. The rejection-slip says, under such a picture: "There isn't even a human being in it to relieve the severity of the building's hard lines and the flat expanse of water. We do not care for such pictures." True, a photograph of a building—and of a building only—may sell for a few dollars to an architectural magazine; but more dollars and a bigger future come from putting life into photographs and in getting your work into the national weeklies as a result.

Again, no magazine wishes to buy a photograph of something not new. A monument, if photographed a moment after the unveiling and with the crowd around it, is a likely seller; but if the photographer waits several years, a print of the monument is unsalable. And that is not strange: you prefer fresh to cold-storage eggs.

The big secret of the successful press-photographer is the introduction of human beings into his photographs of inanimate objects. Human beings have a deep interest in each other. When one is introduced into a picture, human-interest is introduced at the same time; and, if the human being is pictured in the act of doing something, the interest is even higher. For no one ever outgrows the question, "What ya doin', mister?"

Popular Science Monthly says: "We want good, clear photographs of a human being doing something of a mechanical nature. The subjects must be new."

If a new invention is pictured alone, it is lifeless and meaningless. But let a human being operate it and a photograph of it gains in value.

One has only to apply his common sense to the matter. If a murder is committed in the city, the newspapers will not demand photographs of the corpse; it will do very well to obtain a photograph of the "arrow-points-to-the-scene-of-the-crime" variety.

One has to depend wholly on his "nose for news" and this sometimes proves treacherous. "A human-interest photograph sometimes slips past the trained nose of a photographer of twenty years' experience and is picked up by a beginner," to paraphrase Charles Phelps Cushing. And, on the other hand, the old-timer may snap away confidently at a subject which the beginner has scorned, and then find he has an unsalable print on his hands. Sometimes, so to say, "noses for news" contract colds and are unable to scent a subject's salability. But colds may be cured and the scents picked up once more. The best remedy is to stop, to think, and to sniff again.

There is a market somewhere for every good print. There is no market anywhere for a print that is not good.

The best part of the whole business is this: no one—not even old Nick himself—can induce an editor to buy a photograph he does not want; and if, on the other hand, he knows he can use it, he will buy it at once, be it offered by Donald Thompson, who is a world-famed press-photographer, or by John Brown of Smithville, whose first attempt it may be.

SIZE, SHAPE AND FORM

Aspiring fictionists learn at some stage of their budding genius that one long stride toward editorial favor lies in the proper preparation of the manuscript. Just so, a photograph which is not prepared in accordance with editorial standards suffers a handicap.

Some editors specify the size of photograph they prefer. Thus, *Collier's* prefers 4 × 5 prints; but it will use prints larger, and a few smaller than that size. In the same way, *Garden Magazine* reports that it prefers 6½ × 8½ prints, and the Thompson Art Company says it prefers the 5 × 7 or 8 × 10 size.

Other magazines make no mention of size. *Popular Mechanics* reports: "The size of the print is not so important as clearness and gloss." Indeed, the greater number of magazines do not specify a preferable size because by so doing they discourage contributors of prints which are desirable, but not of the size specified.

If a magazine insists on having prints of one certain size the photographer should not be discouraged because his camera does not make photographs of those dimensions. The making of enlargements is now no more difficult than the making of contact-prints; if the negative is sharply focused and the lens of the enlarging-machine is good, an enlargement will not differ much in quality from a small print.

To me, it seems that the ideal camera makes photographs of 3¼ x 4¼ inches. This is very slightly smaller than 4 × 5, and a less costly "film-eater." Negatives of that size are sufficiently large to make salable prints without enlarging them, and if a larger print is desired, they are of good proportions for the operation of enlarging. Prints of the 2¼ × 3¼ size are too small to offer to magazines unless the subjects are all-commanding; however, the size is a very good one, and not too small for the making of excellent enlargements if the lens of the camera is good. I have heard of one photographer who uses exclusively a vest-pocket camera equipped with a fast anastigmat lens: he never attempts to market any of the small prints, whose size is 1-5/8 × 2½, but enlarges the prints to about 4 × 6. There are many advantages possessed by the small camera over the large camera; but 3¼ × 4¼ is the happy medium. I have never had a print of that size returned because it was too small.

There is no need to limit one's self to the production of prints of only standard dimensions. In the cases of magazines desiring artistic prints, the prints gain materially by trimming them so as to produce a compositional balance of masses. Also, some buyers specify prints of a certain shape for use as covers and headings, to fit frame-cuts and such. These buyers state their specifications, as "prints size 4×6, with the long edges horizontal," or the opposite. It is not necessary to produce prints trimmed to the exact size of the cover, either; all that is necessary is to make the print of the same *proportions* as the cover, and the engraver will enlarge or reduce it to the correct size.

There is one best finish for prints intended for publication: that is, black-and-white—*never sepia*—and glossy, burnished. Glossy prints are not much more difficult to make than dull-surfaced prints, the only necessary additional effort being the use of a squeegee plate, or ferrotype plate. The preference for glossy prints results from the fact that their surfaces are absolutely smooth and without grain. This enables the engraver to make a clearer halftone, for a print with a grained surface reproduces surface and all in the cut.

Glossy paper, when dried in the ordinary way, has a surface which is perfectly smooth, yet half-dull. When glossy prints are dried in contact with a ferrotype plate the surfaces are highly polished, and this gives the prints more brilliancy. Prints so prepared are ideal for reproduction-purposes.

Newspapers, as well as some moderate-priced magazines printed on news-print paper, and printed at high speed, require coarse-screened cuts; in these, fancy lighting is detrimental, and fine details are lost; what is wanted are broad masses of light and shade.

Some editors prefer prints which are untrimmed and printed to the very edges of the negative. Such prints give the editor opportunities to trim the prints as he pleases. And in the case of simple news-photographs and ones which have no claim to artistic consideration, it seems to be the preferable method of submission. Certainly, editors will not object to such prints, and they may welcome them in preference to trimmed ones.

Single-weight paper is always preferable to double-weight, even in the larger sizes.

Prints must be sharply focused and distinct—not "fuzzy."

A contrasty print is sometimes recommended as the best to offer; but that is a mistake. The photo-engraver wants prints with plenty of detail in the shadows, and with a tendency to softness; but with not a vestige of flatness. "In the making of the screen-negative and in the various steps of etching, he—the engraver—can introduce highlights into a rather soft subject; but he cannot produce detail in harsh lights and shadows," declares *Photo-Era Magazine.* The process of halftone-making has developed so that the reproduction can be made almost indistinguishable from the original. In any event, make the best print possible—a normal and truthful representation.

Having produced your print, add your name and address to the back of it, and then write, in pencil and on a hard surface, the caption that should be placed under the photograph when it is printed.

Some editors decry the practice of writing the caption on the back of the print; for the print goes to the engraver and the copy for the caption goes to the printer. The alternative is to write the caption on a slip of paper which should be pasted by one end to the back of the print. In any case the photographer's name and address should be stamped on the back.

An ideal print for reproduction and publication, then, should be:

Not smaller than $3\frac{1}{4} \times 4\frac{1}{4}$ inches; on single-weight glossy paper, burnished; very sharp; not contrasty or flat; correct proportions if necessary; untrimmed, if preferred; name and address on back; caption plainly written on back, or on an attached slip.

Prints passing this examination are ready to be shipped to market.

WHERE TO SELL

Once upon a time a publisher had a remarkable inspiration. He would publish a perfect book. He went about the task with painful care. Months were consumed in the making of a book which would be perfect from every viewpoint. After the publisher had corrected every typographical error, had made every possible improvement, and was unable to detect even one flaw in it, he made proof-copies of it and sent them to men on the faculties of universities, to leading printers, to book-making experts, to authorities in English, and to leaders in every other branch of work from which it was possible to view critically the making of the book. He asked them to examine the proofs minutely and to tell him of any flaw, however small, that they might find. Each one of the critics returned his proof with the statement that he had not found the slightest imperfection. Thereupon the beaming bookmaker published his perfect book and offered a large sum to any one who could find a single flaw in it. And many months passed.

Then, one day, he received a letter that pointed out an error in the book. Another letter followed; then another; and at the end of a year, he had received a half dozen letters, each pointing out a different mistake—and each was very noticeably a mistake. And that is the story of the perfect book.

It is with that book in mind that I have decided not to give here the usual list of buyers of photographs. Such a list may be complete and correct when compiled; but by the time it could be put into print and published, lo! some of the magazines would have suspended publication, other new ones would have sprung up, other buyers would have changed their requirements; so that after a year, the entire list would be useless.

I do not add even a list of non-buyers who were once buyers, for the reason that some of them may become buyers again at any moment. Consequently, in my opinion, to place a list of photograph-buyers in this article would be to waste much space, and with the possibility of inconveniencing any photographers who might attempt to use the list after a year or so of its publication.

Furthermore, there are magazines and other books issued yearly which are devoted almost exclusively to listing markets for manuscripts and photographs; these are in a position to make changes, additions and withdrawals with each subsequent issue, and so to keep the lists up-to-date and of value.

One such book is, "Where and How to Sell Manuscripts." This book classifies photographic markets separately; and also lists elsewhere many buyers of photographs. In addition, lists are given of newspapers, postcard-

and-calendar-makers, and lists of magazines devoted to the household, agriculture, gardening, juveniles, sports, outdoors, the drama, music, art, the trades, etc., all of which magazines use photographs. The book is published by the Home Correspondence School, Myrick Building, Springfield, Massachusetts.

Another such book, which is very similar and which contains such lists, is "1001 Places to Sell Manuscripts," published by James Knapp Reeve, at Franklin, Ohio. These are the only two market-books which are enabled to keep their lists up-to-date and correct.

Writer-craft magazines, which maintain literary-market news-columns, list markets for photographs; these supplement the market-books.

The Editor, published weekly at Book Hill, Highland Falls, New York, publishes perhaps more market-notes than any other.

The Writer's Digest, 15-27 West Sixth Street, Cincinnati, Ohio, is a monthly writer-craft magazine which conducts a very good department of market-notes.

The Writer's Monthly is the name of another magazine that lists such markets. It is published monthly. Its market-news, upon publication, is rather older I have found, than that printed in *The Editor*. The longer time necessary to print the magazine may account for that. This magazine is published by the Home Correspondence School, Springfield, Massachusetts.

The Student Writer, 1835 Champa Street, Denver, Colorado, published monthly, maintains an excellent market-list. Their notes are many, varied, and reliable.

Photographic magazines sometimes list markets for photographs, although not frequently.

American Photography, 428 Newbury Street, Boston, Massachusetts, sometimes publishes market-notices in its "The Market-Place" department, but they are scanty.

Photo-Era Magazine lists, when available, market-notes. Book-publishers wishing prints of special character have used this magazine as an advertising-medium.

Besides the magazines noted, other writer-craft and photographic publications may publish market-notes from time to time.

It is by no means necessary to buy both books and to subscribe for all the magazines; but if you can do so without financial discomfort, it cannot be otherwise than to your advantage. By all means, obtain one of the market-books and subscribe for one of the writer-craft magazines; and if you can add a photographic publication, so much the better. Even a market-book alone is a great aid; indeed, it is a necessity. Obtain one or both and you will be amazed at the number of times each can say, "Open Sesame" without stuttering.

The best salesman in the world could not induce a sane blacksmith to put in a stock of groceries. If the salesman has groceries to sell, he goes to a grocer and talks. Similarly, a photographer cannot hope to sell the most remarkable photograph in the world, unless he sends it to the right market.

Each magazine has its own particular needs; but the needs of different ones overlap so far, and are sometimes so similar, that a print offered to one and rejected by it may be very desirable to another; this applies to *classes* of magazines as well as *individual* publications. As an instance: *Popular Mechanics*, or *Illustrated World*, although requiring unusual photographs, rarely buy photographs of human freaks—but nevertheless the *Saturday Blade* (Chicago) uses just that sort of thing.

A few blocks from here stands the largest writing-tablet factory in the world: a photograph of it would not be acceptable to the rotogravure-sections nor to *Popular Mechanics, Illustrated World*, nor to *Popular Science*; yet such a photograph would be useful to an architectural magazine, a stationers' publication, or a local newspaper. When a photograph may be viewed from several industrial angles, as well as from a new-achievement or from a human-interest standpoint, the more likely are markets to open for it. *The press-photographer should not stop until he has tried every possible market.*

After one or two rejections, the photographer is apt to form the opinion that editors are prejudiced against his work because he is a beginner; but nothing could be further from the fact. One national magazine says; "Should we return what you submit, do not be discouraged. Sooner or later, if you study our needs carefully, you will succeed in finding what we are after." The same thing is true of every other magazine. There is not one of them but is eager to buy your wares if you offer them the kind of goods they want.

A rejection is not a rebuke. It is a challenge. It means that your "nose for news" has failed you—has played you false; or that you have tried to sell groceries to a blacksmith. Rest assured that no editor will willfully refuse to accept, pay for and print any photograph which possesses enough merit to warrant acceptance. The editor holds his chair only so long as he produces the kind and quality of magazine its owners want him to produce; and he can do that only by co-operation with contributors. Without contributors he is at sea in a tub. The editor is the best friend the press-photographer can have.

It matters not how much "pull" you have with an editor, or how near a relative you are, or how good a friend, you can't sell a photograph to him unless you "deliver the goods."

Elliot Walker observes: "The way to sell is to give editors what they want and in the way they want it." If you do that you can't fail if you try.

Nor will any editor reject your photographs because of his personal feelings. "The magazine-editor, in the first place, keeps his personal feelings tied up; in the second place, he would be foolish, indeed, to allow them to influence his decisions; and, in the third place, the editor 'ain't got no' personal feelings when it comes to buying material for his magazine."

There is only one course to pursue—send the photograph to every possible market for it in its special line; then see if it can be viewed from another magazine-angle, and try every magazine of that trend; then repeat and repeat and ship it away again and again. *Don't stop until it has been returned from every market with the slightest possibility of buying it.* Then sit up nights to discover another shipping-point for it. Keep on to the bitter end; but if your "nose" is working and you keep on steadily, the end will come rather suddenly, and it will not be bitter.

A SURVEY OF MARKETS

What follows is no attempt to list and classify existing markets, but to offer a generalized survey of magazine needs by class. While the success of the small-town press-photographer is not in proportion to his city's size, the magazines which find their ways to him month after month do not disclose the whole field of markets to him. He needs something more—something to reveal to him the broad needs of magazines. This chapter has as its mission the summarizing of the needs of magazines of every class.

Thus, photographs taken all over the world, showing the beauty and commerce of the old and new eras, are eagerly sought by several magazines. *Travel*, 7 West Sixteenth Street, New York, wants photographs of out-of-the-way places, unusual methods of producing world necessities, and photographs of general travel interest.

The same may be said of the *National Geographic Magazine*, though the photographs and articles used by this publication are so specialized and exhaustive that it is rarely a free-lance writer can supply their needs—for they maintain their own staff of writers and explorers. However, if you are able to catch vivid photographs of wide travel interest, here is a most excellent market.

If you are interested in picturing homes, *Country Life*, *Garden Magazine* and *House Beautiful* are waiting for your prints. These magazines are very artistic and use only the best work; but they are interested in unusual gardens, beautiful lawns, landscaping, interior decorating. A house remodelled from a common building to an unusual or striking residence will find ready sale to them if photographs of the "before and after" variety are offered. Nature, sport, and building in the country are the specialty of *Country Life*, Garden City, New York; *Garden Magazine* is interested in nothing but gardens and ornamental horticulture, preferably of the personal experience trend. Same address as *Country Life*. *House Beautiful*, 3 Park Street, Boston, wants photographs of unusual types of interior decorating and landscape architecture. What a wealth of material a well-kept, modern home contains! Owners should readily give consent to photograph if the photographer explains his purpose.

Arts and Decoration, 470 Fourth Avenue, New York, also uses garden and house material, but runs also to the arts. Photographs of architecture, interior decorating, etc., here find another market.

So it is with the broad field of country-life magazines generally, as an example. House furnishing and "before and after" remodelling pictures are easily obtained and easily sold if well done.

Every class of magazines uses photographs: Literary magazines, Women's, Farm journals, Juvenile, Religious, Outdoor, Photographic, Theatrical, Musical, Art, and Trade publications. The following notes generalize the needs of each of these fields.

GENERAL MAGAZINES

This excludes most fiction magazines; those which do use photographic illustrations buy the work of professional studios already established and perhaps specializing in that type of illustrating. The beginner may develop into one of these illustrators—many magazines use them, as *Love Stories*, *Cosmopolitan* for special articles, *National Pictorial Monthly*, etc.,—but these markets are not open to the free-lance photographer.

Current History, Times Building, New York, New York, is an example of a news-magazine which uses timely photographs of wide interest.

The Literary Digest is of similar nature, but this second magazine does not buy photographs from the open market.

The Curtis Publishing Company occasionally uses photographs of a scenic or artistic nature as fillers. These magazines comprise *The Saturday Evening Post*, *The Ladies' Home Journal*, *The Country Gentleman*. These are always available, and a glance through several numbers of each will disclose the type of photograph wanted.

Grit, Williamsport, Pennsylvania, uses many photographs, and short articles written around them. This publication wants common, human-interest subjects treated carefully.

The needs of *The Illustrated World*, *Popular Mechanics* and *Popular Science* have been made very clear in previous portions of this book.

The Scientific American always wants photographs of new inventions of wide interest, accompanied by brief articles. Address 233 Broadway, New York, New York.

Physical Culture, 119 West 40th Street, New York, New York, always wants photographs of persons having splendid physical development. A glance through this magazine will disclose the types of poses desired. Straight front, back, etc., views are never used; action in the picture is essential.

WOMEN'S MAGAZINES

These magazines use generally pictures of home improvements, remodelling of residences, flower gardens of unusual variety, and use short illustrated articles on house-building, interior decoration, rugs, gardens, domestic science, etc. The magazines listed below are only a few of the many which use photographs and illustrated articles of interest to women.

The Ladies' Home Journal, Philadelphia, Pa.; the *Woman's Home Companion*, New York; the *Delineator*, New York, and *Good Housekeeping*, New York, are all generally fiction magazines with a homey flavor which do not offer a good market for separate photographs or short illustrated articles, although they are in the market for suitable material of this sort, in a limited way. Others are:

American Cookery, 221 Columbia Ave., Boston.

Better Times, 70 Fifth Ave., New York.

Canadian Home Journal, 71 Richmond St., West, Toronto, Ontario, Canada.

Farm and Home, Springfield, Mass.

Mother's Magazine, 180 No. Wabash Ave., Chicago.

New England Homestead, Springfield, Mass.

Vogue, 19 West 44th St., New York, uses exclusive photographs of society in New York, Newport, etc.; photographs of handsome homes of well-known society people, beautiful and unusual gardens, etc.

Woman's Weekly, 431 So. Dearborn St., Chicago, uses short articles of home interest, illustrated.

FARM JOURNALS

The needs of farm journals are specific. They form an important division of published magazines, and a large one which uses a great amount of material. Articles on farm improvements, etc., are always used, and photographs also. A conjunction of the two, in an illustrated article, forms a much more marketable commodity. The farm work is composed of many divisions— agriculture, bee culture, botany, breeding, cheese-making, etc. The following are a representative few of the agricultural markets which are always buying material:

American Agriculturist, 315 Fourth Ave., New York.

American Bee Journal, Hamilton, Ill.

American Botanist, Joliet, Ill.

American Breeder, 225 West 12th St., Kansas City, Mo.

American Farming, 537 So. Dearborn St., Chicago.

American Forestry, 1410 H St., Washington, D.C.

American Fruit Grower, State Lake Bldg., Chicago.

American Poultry Journal, 542 So. Dearborn St., Chicago.

American Seedsman, Chicago, Ill.

Bean-Bag, Syndicate Trust Bldg., St. Louis, Mo., is devoted to the bean industry.

Canadian Countryman, 154 Simcoe St., Toronto, Ontario, Canada; material of Canadian interest.

Country Gentleman, Independence Square, Philadelphia.

Dairy Farmer, Waterloo, Iowa.

Farm and Fireside, 381 Fourth Ave., New York.

Farm Journal, Philadelphia, Pa.

The Horse World, 1028-30 Marine Bldg., Buffalo, New York.

Jewish Farmer, 174 Second Ave., New York.

Kennel Advocate, 636 Market St., Sierra Madre, Cal.

The Milk Magazine, Waterloo, Iowa.

National Alfalfa Journal, Otis Building, Chicago.

Orchard and Farm, 1111 So. Broadway, Los Angeles, Cal.

Potato Magazine, Room 605, 139 No. Clark St., Chicago.

Power Farming, St. Joseph, Mich.

Rabbitcraft and Small Stock Journal, Lamoni, Iowa.

Southern Agriculturist, Nashville, Tenn.

Wallace's Farmer, Des Moines, Iowa.

JUVENILE PUBLICATIONS

Almost every magazine uses juvenile material, and there are many that specialize in it. The following markets use the well-known type of photograph and illustrated article which are of interest—travel, how-to-make-it, etc. A great field is open here to picturized activities of boys.

The American Boy, 142 Lafayette Blvd., Detroit, Mich.

Boy's Magazine, Scarsdale, N.Y.

Classmate, 420 Plum St., Cincinnati, Ohio.

Forward, Witherspoon Bldg., Philadelphia.

Girl's World, 1701 Chestnut St., Philadelphia.

Junior Christian Endeavor World, 31 Mt. Vernon St., Boston, Mass.

Kind Words, Nashville, Tenn.

Open Road, 248 Boylston St., Boston.

St. Nicholas Magazine, 353 Fourth Ave., New York.

Youth's Companion, 881 Commonwealth Ave., Boston, Mass.

RELIGIOUS PAPERS

Religious publications are not given to printing many photographs, although there is a market of appreciable size here. This field is a difficult one to generalize upon, but the following may be taken as such a list:

Adult Student, Nashville, Tenn.

American Messenger, 101 Park Avenue, New York, New York.

Christian Advocate, 810 Broadway, Nashville, Tenn.

Christian Endeavor World, 31 Mt. Vernon St., Boston, Mass., uses photographic covers.

David C. Cook Company, Elgin, Illinois, publishes about forty magazines, which use a great amount of photographs and illustrated material.

Epworth Herald, 740 Rush St., Chicago.

Front Rank, 2710 Pine St., St. Louis, Mo.

Lookout, Cincinnati, Ohio, uses photographs for covers.

The Missionary, Apostolic Mission House, Brookland, Washington, D.C.

Sunday School World, 1816 Chestnut St., Philadelphia.

The Watchword, Otterbein Press, Dayton, Ohio.

OUTDOOR MAGAZINES

Here is a group of magazines which is deeply interested in unusual fishing-trips, hunts, and such excursions—it wants material on the animals in water or air or on land, that its readers may bag them the more easily; it desires material on bird-dogs, on outdoor devices and tricks, on tennis, motoring, baseball, cats, dogs, golf, horses, yachting, and on every phase of outdoor and sport life. Photographs of men prominent in each line are wanted; prints of hunting, fishing, camping, canoeing, sailing, and everything connected with the big outdoors. Here is a large and remunerative market for open-air photographs and sport prints.

Aerial Age, 280 Madison Ave., New York, wants material on aviation.

All Outdoors, Outing, Forest and Stream, Field and Stream, etc., want the wide variety of outdoor material that appeals to any sort of sportsman. These magazines circulate widely, and a study of them will disclose their needs.

Dogs are the subjects of such magazines as *American Beagle*, 639 West Federal St., Youngstown, Ohio; *Dogdom*, Battle Creek, Michigan; *Dog Fancier*, Battle Creek, Michigan; *Dog World*, 1333 So. California Ave., Chicago.

Material about cats is welcomed by such as *Cat Review*, 196 Centre St., Orange, New Jersey.

Fishing material appeals to the general run of outdoor magazines, including *American Angler*, 1400 Broadway, New York.

Tennis appeals to *American Lawn Tennis*, 120 Broadway, New York, and the *Tennis Review*, California Bldg., Los Angeles, Cal.

Golf material is used by *American Golfer*, 49 Liberty St., New York, and *Golfer's Magazine*, 1355 Monadnock Block, Chicago.

Motoring appeals to a long list of such publications as:

American Motorist, Riggs Building, Washington, D.C.

Mileage, 4415 No. Racine Ave., Chicago.

Motor, 119 West 40th St., New York.

Motordom, 110 State St., Chicago.

Motor Life, 239 West 39th St., New York.

Speed, 809 Shipley St., Wilmington, Del.

Then there are a variety of different subdivisions of this class, the mere names of which are sufficient to disclose the great variety of material they use:

American Checkers, 1846 So. 40th Ave., Chicago.

American Chess Bulletin, 150 Nassau St., New York.

American Cricketer, Morris Building, Philadelphia.

Baseball Magazine, 70 Fifth Ave., New York.

Billiards Magazine, 35 So. Dearborn St., Chicago.

Bird Lore, 29 West 32d St., New York.

Bowler's Journal, 836 Exchange Ave., Chicago.

The Horse World, 1028-30 Marine Bank Bldg., Buffalo, New York.

Spur, 389 Fifth Ave., New York—raising prize winners.

Yachting, 141 West 36th St., New York.

PHOTOGRAPHIC MAGAZINES

These magazines pay more attention to the photograph itself than to what it pictures. Here is a market for artistic prints, for prints showing new working methods, and such material interesting to photographers. Artistic taste and technical accuracy are instrumental in getting you into these magazines.

American Photography, 428 Newbury Street, Boston.

The Camera, 210 No. 13th St., Philadelphia, Pa.

Camera Craft, Claus Spreckels Bldg., San Francisco, Cal.

Photo-Era Magazine, Wolfeboro, New Hampshire.

THEATRICAL MAGAZINES

Theatrical magazines embrace the following representative few who desire prints of current news in the show world, new theatres, interviews with actors and actresses and photographs of them, etc.

The Drama, 306 Riggs Bldg., Washington, D.C.

Theatre Arts Magazine, 7 East 42d St., Detroit, Mich.

Theatre Magazine, 6 East 39th St., New York.

MUSICAL JOURNALS

Photographs of bands, orchestras, leaders, band-stands that are unique, artists, composers, etc., are used by this class.

Musical Courier, 437 Fifth Ave., New York, New York.

Musical Enterprise, Camden, N.J.

These include magazines published and devoted to every trade imaginable. One magazine will be cited for each division of trade, the title of which is self-explanatory, and which uses photographs in its particular field:

Advertising: *Advertising and Selling*, 471 Fifth Avenue, New York.

Architectural: *American Builder*, 1827 Prairie Ave., Chicago.

Automobile: *American Garage and Auto Dealer*, 116 So. Michigan Ave., Chicago.

Baking and Confectionery: *Baker's Helper*, 327 So. La Salle St., Chicago. *Western Confectioner*, Underwood Bldg., San Francisco.

Cement, etc.: *Concrete*, 314 New Telegraph Bldg., Detroit, Mich.

Drug, Oil, Paint, etc.: *Druggists' Circular*, 100 William St., New York. *Painters' Magazine*, same address.

Dry Goods: *Dry Goods Reporter*, 215 So. Market St., Chicago, Ill.

Electric: *Journal of Electricity*, Crossley Bldg., San Francisco.

Engineering: *Everyday Engineering Magazine*, 2 West 45th St., New York.

Financial: *Financial World*, 29 Broadway, New York.

Fraternal: See particular paper referring to particular fraternity or lodge in list given in Market Book.

Furniture: *Furniture News*, Wainwright Bldg., St. Louis, Mo.

Grain: *Grain Dealers' Journal*, 315 S. La Salle St., Chicago.

Grocery: *National Grocer*, 208 So. La Salle St., Chicago.

Hardware: *Good Hardware*, 211 So. Dithridge St., Pittsburgh.

History: *Hispanic American Historical Review*, 1422 Irving St., N.E., Washington, D.C.

House Organs: Some two thousand of these are listed in the market books named.

Jewelry: *Jewelers' Circular*, 11 John St., New York.

Labor: See particular division desired by consulting Market Book.

Law: *Casualty Review*, 222 East Ohio St., Indianapolis, Ind.

Lumber: *Lumber*, Wright Bldg., St. Louis, Mo.

Medical: See division desired, as Dental, Hospital, etc., in Market Book.

Military: *American Legion Weekly*, 627 West 43d St., New York.

Municipal: *American City*, 87 Nassau St., New York.

Printing: *The Inland Printer*, Inland Printing Co., 632 Sherman St., Chicago.

Railroad: *The Railroad Red Book*, 2019 Stout St., Denver, Colo.

Shoes: *Boot and Shoe Recorder*, 207 South St., Boston.

This survey gives a general idea of the wide market open to photographs which fall within each magazine's requirements. No attempt has been made to give the needs of magazines, or to present what is usually called "a list of markets." We have been concerned here with generalizing the market— pointing out to the reader who never sees most of the magazines named that they really exist and buy photographs. The purchase of a Market Book is necessary if one desires seriously to make his way selling photographs to publications.

"Study the magazine" is the bromide flung always in the teeth of the beginner. But what if one can't obtain copies of the magazines which print material which the reader may easily find? Then he has only to request from the editor a sample copy of the magazine, using the address gleaned from the Market Book—and he then has the best information as to what that particular magazine wants. And at a cost of only two cents per copy.

SHIPPING THE PRODUCT TO MARKET

When a print is to be offered to a local newspaper, the photographer starts out, sometimes, as soon as one hour after making the exposure, with the print in his hand, and, arriving at the desk of the city-editor, he allows him to examine it. In such a case, mailing the print would delay it; perhaps delay it until its interest has cooled, and so make it worthless. But when submitting prints to magazines one should always invoke the aid of Uncle Sam's mail-service, no matter if the editor lives just next door and the publication-office is but a block distant.

The shipping of your prints to their markets merits special consideration. If the photograph, after being wrapped, can be bent easily, it is apt to arrive at the editor's desk in a cracked and crumpled condition. Then the editor could not buy it if he wished. And, when it is returned, its maker finds it to be so mutilated that it is useless to try to market it elsewhere. Proper protection of photographs when shipping them is an aid to both editor and contributor.

Photographs which are 4 × 5 inches in size can be sent safely in a No. 11 envelope of heavy manila paper if a sheet of cardboard is placed in the envelope too. The cardboard prevents the breaking of corners, the bending, and the cracking of the print. For a return-envelope—*never omit to enclose an envelope addressed to yourself and adequately stamped for the return of the print if it is unavailable*—for a return envelope, a No. 10 manila envelope is the best.

Prints which are 4 × 5 inches or larger should be sent in larger envelopes—in clasp-envelopes. These envelopes can be obtained at stationers' in sizes suitable for almost any photograph. The envelope should be about an inch larger each way than the print. The print, as well as a piece of cardboard—which should be somewhat larger than the print—can be sent safely in the clasp-envelope container. *On no occasion forget to enclose a return-envelope, which should be self-addressed and stamped.* The return-envelope may be of the same size as the outer one; and, if it is folded, it may be easily inserted. The envelopes mentioned, I have found by experience, are the best containers that can be used for photographs that are to be mailed.

Never roll a print and insert it in a mailing-tube. If there is anything an editor does *not* want you to do, it is that. Prints so sent never lose the violent curve they acquire in transit, and then they are no more amenable to reason than a temperamental mule. Prints should always be sent *flat*—never rolled or folded, nor in any other condition except perfectly *flat*.

The envelope should be addressed to "The Editor" of the particular magazine selected. Do not address it to the editor by name, for it might arrive at a time when he is on his vacation, and so it will follow him all over the country and perhaps become lost. There should be no enclosure other than the photograph; except, when it is necessary, a sheet carrying an explanation or a short article to be printed with the picture. Do not write a letter to the editor unless the photograph is timely and should have an immediate decision. The professional news-photographer submits his work without letters, and with no identification except his name on the back of each print—and it isn't what's on the back, but what's on the front, that counts.

Photographs properly require only third-class postage rates. The addition of a caption to the print, or any other written matter included with it, automatically raises the rate to first-class. Even if nothing but the photograph alone is sent, I advise the use of first-class service for several reasons: the print is then carried more quickly; it is handled more carefully; and the sender may seal the container, which he is unable to do with third-class matter. Always, then, send your photographs by first-class mail.

Editors do not maintain special funds for the purpose of paying for postage-due stamps. That is, if a package of photographs arrives at the editor's desk with the postage not fully prepaid, the payment by the editor of the postage due does not make his attitude kindly toward the work itself. There are a good many editors who will not accept contributions from the postoffice which have postage-due stamps attached because of the neglect of the sender to fully prepay the postage. There are a great many more editors who will not return photographs unless a stamped and self-addressed envelope is enclosed with the offering. The attitude is entirely justified, for the supplying of postage to careless contributors in such cases would cost a magazine hundreds of dollars every year.

Never send your photographs by registered mail unless their value is extraordinary; and never send them by special-delivery mail unless the prints are addressed to a newspaper and possess burning-hot news interest. To send photographs of average quality by either registered or special-delivery mail is a trick of the novice struggling for recognition. Use ordinary first-class service and the editor will feel more kindly toward you than if he is made to stop his work and sign a mail-receipt.

Not all photographs are accepted by the very first editor who sees them. Very often it is the fifth, or the tenth, or even the twentieth editor who buys them. So if a print comes back, immediately send it out again and again and again. *Don't stop, for the very next time you might sell it.* If it's a good print, there is an editor somewhere waiting for it.

THE PRICES PAID

The most remarkable news-photographs ever made—they were exposed at the South Pole—brought $3,000 from *Leslie's* (now no longer published) for "First Rights," and $1,000 more from International Feature Service for "Second Rights." Some photographers have realized hundreds of dollars from lucky shots; an extraordinary photograph may bring from $25 to $100; but the average price paid is $3.00; and, indeed, there are some editors who unblushingly offer as little as ten or twenty-five cents for prints; and some who find it impossible, unwise, or unnecessary to pay for prints at all.

Although the average price paid is not astounding, it is a good return on the cost of making; also, the abundant opportunities for salable prints compensate for what each cheque lacks. A photographer who is wide-awake and moving ought not to find it difficult to sell at least ten prints each week, if not more, when one considers the large number of available subjects and the multitude of magazines.

Newspapers pay for prints according to their breadth of circulation. A widely-read daily will pay more for photographs than one of small circulation. Very often, newspaper-editors prefer that the press-photographer send a bill for his services. If you are asked to do that, do not hesitate to charge a price you think is entirely just; but don't grasp the opportunity to profiteer. Better, discover the price asked by the newspaper's favorite commercial-photographer, and mark down your price accordingly. That is business; it isn't taking an unfair advantage.

Whatever the price that is paid, don't object if you think it is too low; accept the payment and seek a more remunerative market next time. This applies to magazines as well as to newspapers.

The prices paid by magazines vary likewise, but none of any reputation pays less than one dollar per print. There are many factors which decide the size of the cheque which the press-photographer receives. The first is the circulation of the publication, for its financial reserve depends on the number of buyers. The size of the print in some instances decides the price paid. Thus, one magazine pays $1.00 for prints of one size and $2.00 for larger ones. However, there are not many magazines who pay according to the size of print.

Sometimes, retouching must be applied to a print in order to make it suitable for reproduction; and, as the service of a retoucher is expensive, something is deducted from the photographer's cheque to pay for the work. *Popular Science* is a magazine of that policy. The photographer can avoid such deductions from his cheques by supplying photographs of such quality that they will need no retouching.

If a photograph is offered for the exclusive use of one magazine it may bring a higher price than if it were non-exclusive. Thus, *Collier's* pays $3.00 for non-exclusive prints and $5.00 for exclusive ones. Some few magazines rarely accept any print that is not exclusive; indeed, non-exclusiveness may be a reason for rejection. Calendar-makers and postcard-makers, of course, buy only exclusive rights. A publisher is always more favorably inclined toward an exclusive than toward a non-exclusive print; and, very often, the added favor means added dollars to the payment.

The use to which a print is put is also a deciding factor in payment. A print bought for use as a cover-illustration will bring home a bigger cheque than if it were used merely as one of many illustrations. Too, *Illustrated World* pays $3.00 and more for prints used in its pictorial section, but $2.00 for those used in its mechanical department. Other magazines do not make this distinction.

After all, the price paid depends wholly on the usefulness and quality of the print. If, sometimes, as in the case of the *Ladies' Home Journal*, the payment is made with a view to the photographer's reputation, it is only because news-photographers of experience produce prints of a higher average quality than beginners do. But, if a beginner "delivers the goods," the editor is just as glad to pay to him the large cheque as he is to pay it to any one else.

A few examples of prices paid will be of interest. *Collier's* pays $3.00 for non-exclusive prints and $5.00 for exclusive prints, and from $25.00 to $100.00 a page for layouts (spreads). *Illustrated World* pays $3.00 for each print. *Popular Mechanics* pays $3.00 and up, and $25.00 a page for layouts. *Popular Science* reimburses at the rate of $3.00 for each photograph, and sometimes more. The *Saturday Blade* pays $2.00 for each. The Thompson Art Company pays from $1.00 to $5.00. Underwood and Underwood pay from $3.00 and up, according to the value of the print. The Woodman and Teirman Printing Company pays at rates varying from $5.00 to $50.00.

"But when is payment made?" you ask. The answer is, "Either upon acceptance or upon publication."

By far, most magazines pay according to the more desirable plan—upon acceptance. As soon as such a magazine decides that a photograph is useful to it, it mails a cheque to the sender. Sometimes, a receipt is sent with the cheque, which the recipient must sign and return; but, more often, the cheque itself is the receipt. Payment upon acceptance is by far the more desirable method, for with it the worker is paid as soon as his work is done;

there is no waiting for weeks and months for payment, as in the case of pay-on-publication magazines.

There are a few magazines who wait until the photograph actually appears in the pages of the publication before payment is made. In such cases, the photographer has no recourse but to wait until the editor is ready to print his contribution whenever it may be.

In the case of pay-on-publication magazines, notice is usually sent that the photograph has been accepted for publication and that it will be paid for as soon as it is published. Sometimes, no notice is given at all of publication or acceptance; and in that case the photographer must scan each issue of the magazine in order to find his contribution when it appears, or he must wait until the cheque arrives that denotes publication. Either method is uncertain; but there is nothing to do but to endure it. Some publications even wait for some time after publication before making payment, as in the case of the *Kansas City Star*, which pays on the fifteenth of the month following publication, and the *Saturday Blade* which also mails all cheques the month following publication. This is a discouraging policy; but as the cheque always arrives in the end, there is little to be said in condemnation of it; the photographer is obliged to make the best of it.

The contributor should always keep a record of prints accepted and to be paid for on publication. Otherwise, by an oversight, a cheque for published material may never come, and the photographer may never miss it. Too, a cheque may arrive unexpectedly from a forgotten source and cause an attack of heart-failure.

The beginner does not achieve mountain-top prices except by a lucky shot now and then. Prices increase with your experience and your reputation.

The photographer who develops his "nose for news" until it can scent a salable photograph in every conceivable situation is the photographer who has the large cheques forced upon him.

The sky-high cheques come to the camerist who, night and day, through sunshine and storm, earthquake and cyclone, is always "hot on the trail" of the salable photograph that is tucked away somewhere, where only a keen scent and a large amount of perseverance can lead him; and when he arrives, the subject will be singing truthfully, "Shoot me and the wor-rld is tha-hine." There are enough of these subjects to shame the biggest choir on earth by their "singing." However, the photographer must know good music when he hears it.

ART PHOTOGRAPHS

An art-photograph may be either of two things: a photograph, itself artistic; or a photograph of some artistic thing. There are markets for both. Artistic photographs are used by calendar and postcard makers; also, by photographic magazines, and magazines given to the beautiful in art or literature. When submitting such photographs to makers of postcards and such, they should be submitted in the usual manner.

The subjects used by card- and calendar-makers are interesting landscapes, beautiful seascapes, pretty girls, attractive children, and animals, as every one knows. Such pictures are sometimes bought outright—indeed, they usually are; but some firms pay according to their value as indicated by the demand for them after publication. Thus, one firm pays on a fifty-fifty basis.

An example of beautiful photography, at the same time picturing an unusual or artistic subject, will usually find a market in a photographic magazine, as *Photo-Era Magazine* or a magazine such as *Shadowland*. The *Architectural Record* demands that its prints, although of architectural subjects, be artistic and beautiful. Indeed, there is such a wide market for photographically artistic prints of beautiful subjects that the photographer is doubly rewarded who can supply these, as well as hot-off-the-bat news-photographs.

Artistic photographs are printed on sensitive-paper of a surface suited to their subjects, and are trimmed so as to carry the correct compositional balance; and after, they are tastefully mounted.

Photographs which are not themselves artistic, but which are of art-subjects, may be prepared as are other photographs intended for publication. Such photographs are of statues, pictures, new art-museums, art-collections, paintings, mural decorations, drawings, and anything at all of interest to artists. Material of such sort is sought by such publications as *American Art News*, *Art in America*, *Art and Decoration*, and others that appreciate the very best.

In short, the photographer may market his game among a wider patronage if he can bring down birds of paradise as well as ducks and geese and the common denizens of the air.

COMPETITIONS

Competition is the life of business. Certainly, then, an aspirant for honors from publishers experiences no lack of life. Often, however, after a print has proved unavailable for publication, when offered by the regular process, it may be entered in a photographic competition where current interest is not essential; and so, perhaps, even bring home a larger cheque than it could have captured otherwise.

The two leading photographic publications, *Photo-Era Magazine* and *American Photography*, conduct monthly competitions. The monthly prizes for the Advanced Competition of *Photo-Era Magazine* are $10.00, $5.00 and $2.50 in value of photographic goods. Although cash is not paid, a prize awarded will go a long way toward obtaining for the photographer a desired piece of apparatus, or in supplying sensitised material, developing-agents and such with which to produce photographs intended for other magazines. "The contest is free and open to photographers of ability and good standing— amateur or professional." The publisher of *Photo-Era Magazine* assigns subjects for each month, as "Winter-Sports," "Speed-Pictures," and so on. Since the photographer must buy supplies in any event, the awarding of such to the amount of $10.00 is a distinct help.

American Photography also conducts monthly photographic contests. For these no subjects are assigned. The prizes for the Senior Class are $10.00, $5.00 and $3.00, paid in cash. "Any photographer, amateur or professional, may compete." This magazine last year held an Annual Competition, which it intends to repeat, with prizes of $100.00, $50.00, two of $25.00, and ten of $10.00, not to mention one hundred subscriptions for the magazine. Highly artistic work is necessary for recognition in the Annual Competition. Both *Photo-Era Magazine* and *American Photography* supply data-blanks which must be sent with entries.

Competitions for amateur photographs are also conducted by the *American Boy*, which offers monthly prizes of $5.00, $3.00 and $1.00 for "the most interesting amateur photographs received during each month." These are worthwhile.

Photographs of popular interest are used in monthly competitions by many magazines; and many manufacturers conduct occasional, if not regular, prize-contests.

Probably the largest company to offer prizes in competitions is the Eastman Kodak Company. The Eastman company for many years conducted a yearly contest with thousands of dollars in prizes offered. Last year, it decided on an innovation; the running of a monthly contest with prizes of $500.00. This practice has been continued for many months and shows no signs of being discontinued at this writing. Prizes are offered for four classes of photographs, the class being determined by the camera with which the photograph was made. In all, twenty prizes are awarded each month, the highest being $100.00 and the lowest $7.00. Frequently one person wins two or three prizes. The photographs entered must be of good workmanship, of human-interest and must preferably tell a story. No subjects are set. Upon writing to the company, a leaflet is sent which gives rules and an entry-blank. A good many photographers have cleaned-up in these competitions.

Now and then, different manufacturers and magazines, who do not ordinarily do so, offer prizes for photographs. At every opportunity, the press-photographer should enter his prints, for if they win a prize, he has the advantage of a larger remuneration as well as a boosted prestige among editors and publishers.

PRINTS FOR ADVERTISING

Advertisers who are manufacturers are all possessed of the belief that the buying public is painfully ill-informed of the unequalled merits of their products. Consequently, any photographic evidence of the superiority of their goods which will enlighten the public is welcomed with open arms.

Any photograph that shows plainly the excellent service that any product has given will bring the photographer's own price from the manufacturer. The demand is almost universal.

Makers of camera-lenses are continually on the lookout for unusual photographs made with their products. The Wollensak, the Bausch and Lomb, and the Goerz companies frequently buy negatives that portray vividly some features of their lenses.

Makers of camera-shutters also buy photographs which were made with cameras equipped with their shutters. Usually, the point emphasised in the pictures bought is the shutters' ability to "stop motion" at their high speeds. As press-photographers frequently find it necessary to use the shortest exposures given by their shutters, they should have something in their negative-files which the shutter-makers should be eager to obtain.

Makers of photographic material other than lenses and shutters often buy examples of work done with their goods. Thus, the Ansco Company "uses photographs of natural scenes for advertising-purposes," the photographs being made on *Ansco* film and *Cyko* paper, or other Ansco products. Burke and James, makers of *Rexo* cameras, "use photographs for advertising-purposes which must be of unusual interest and must illustrate their goods in use, or be made with their cameras or films." Inasmuch as the news-photographer, in his daily work, finds many unusual things, he should find no difficulty in selling a few prints to camera-makers.

An advertiser is always seeking any information likely to help sell his product. If, in your work, you see an old storage-battery with electric energy still unimpaired, or a well-preserved tire, or a shaving-brush of "strong constitution" unweakened by much use, it would very likely prove profitable to photograph it and describe your find to the company that makes the product.

Thus, an insurance-agency may buy a photograph of a garage destroyed by fire, the cars in which were fully protected by their insurance. A maker of strong-boxes may appreciate a photograph of one of his boxes raked out of, perhaps, the same fire, the box having held valuable papers which were fully protected from the terrific heat. The makers of a portable typewriter once purchased a photograph of one of their machines which had fallen from an airplane and which had to be dug from the ground; but which, of course, suffered no injury whatever because of its fall and burial. If you should unexpectedly come upon Irvin Cobb writing a masterpiece with his Neverleek fountain-pen, snap him (with his permission) and see what the makers of Neverleeks say. Manufacturers of patent roofings use photographs of roofs covered with their products; makers of steam-rollers want photographs of roads tamped by their machines; and so on and on and on.

It is wiser to write first to the advertising-manager of the particular company favored, and to inquire if he is buying photographs that show plainly the unparalleled merits of his excellent product, and if so—etc., etc.

Some advertisers will ask you to name a price for your work, and on such an occasion you should judge fairly the value of the print to them. If they require the negative also, raise the rate. Any prints should be worth $10.00 even to a small manufacturer, and if it is acceptable at all, a larger firm should pay from $25.00 to $1,000.00 for suitable propaganda. This branch of press-photography is little used by many workers, yet it is remunerative.

Besides furnishing the manufacturer with advertising for his product, the photographer supplies himself with some advertising to the effect that "he delivered the goods once, and could do it again, so there."

COPYRIGHTS AND OTHER RIGHTS

If, as often happens, one photograph is useful to more than one publication, is it all right to sell the one photograph to as many magazines as will buy it?

When a publication prints a photograph on its pages, it copyrights it in the name of the publishing company. The photographer then has parted with his *entire rights* to it, and cannot sell it elsewhere, *unless* one of two precautions has been taken.

The first precaution is the writing on the back of each print: "First Magazine-Rights Only." Those "mystic" words mean that the print is offered for publication only one time, after which it again becomes the property of the photographer. That is, the magazine, when buying such a print, buys only the right to print it the first time. Immediately after its publication, it becomes again the property of the photographer, although he cannot of course sell "First Rights" again, any more than he can sell the same horse twice at the same time.

After "First Rights" has been sold, the photographer may then sell "Second Rights," *provided* those words are written on the back of the second print. "'Second Rights' is the right to publish a photograph in some other publication than the one in which it originally appeared." For instance: a photograph of a novel shop-window display may be acceptable to *Popular Mechanics*, which buys a print *marked* "First Magazine-Rights Only." But the same photograph may be acceptable too to an advertising-magazine, and so it buys "Second Magazine-Rights." Unless these terms are written on the backs of prints which are sold to more than one magazine, trouble is apt to result.

Another plan by which it is possible to sell a photograph to more than one publication is the labeling *each print* as: "Non-Exclusive" or "Not Exclusive." When that is done, the photograph may be sold to as many editors as care to buy it.

If no mention of any rights or of exclusiveness is made at the time of sale, it is inferred that the publisher buys "All Rights." In that case the photographer loses *all* claims to the photograph; if he attempts to sell it again without the consent of the editor who first bought it he is breaking the copyright laws; in fact, he is selling another's property.

There is no need to affix any such terms to any photograph which can sell to only one, or which is to be offered to only one magazine. Magazines are more partial to prints which they can buy outright, and thus acquire "All Rights." Indeed, there are very few prints of enough value to sell to more than one magazine.

Now we plunge deep into the mysteries of copyrights. When a print is copyrighted it is unalterably the property of the person *first* copyrighting it until he signs "Transference of Copyright." A copyrighted print may be published in a dozen publications if they will buy it, and it still remains the property of the one who first copyrighted it. Copyright laws were passed for the benefit of those who "promote the progress of science and useful arts." This is done "by securing for limited times to authors and inventors the exclusive right to use their respective writings and discoveries." Under this law, "author" includes makers of photographs, and "writings" includes photographs.

The process of copyrighting a photograph is not an involved one. A request should be addressed to the Register of Copyrights at Washington, D.C., for a few copyright-blanks, form J1. (Form J1 is for photographs to be sold, J2 for photographs not to be sold.) One of these cards is then filled out, and two prints of the photographs sent with it to the Copyright Office, as well as the necessary fee. "The fee for the registration of copyrights ... in the case of photographs, when no certificate (of copyright) is demanded is fifty cents; for every certificate, fifty cents" additional. A certificate is not usually necessary, and is useful only in cases of disputed copyright ownership, etc. The fee should be sent only in the form of a money-order to the Register of Copyrights, and the photographs must bear the mark of copyright, which is "either the word 'Copyrighted' or the abbreviation 'Copr.' accompanied by the name of the copyright proprietor. In the case of photographs the notice may consist of the letter C inclosed in a circle *provided* that on some accessible portion of such copies ... the name of the person copyrighting shall appear." Upon the Copyright Office receiving the photographs, the sender is notified; and again, when copyright is granted, he is sent a small card notifying him, or the certificate is sent to him if he has ordered one. Then the print is considered copyrighted.

It is useless to copyright any except those prints of extraordinary value, the rights of which the photographer wishes to retain at all costs. The average quality prints are not likely to be stolen, and so the copyrighting of them is unnecessary. If the photograph is merely to be offered to two or more publications it is only necessary to mark each print as directed in the foregoing paragraphs.

Publishing companies are business-institutions which are of necessity conducted according to the highest ethics. To unwittingly sell to another magazine a print one magazine purchased as exclusive, would be likely to exile the photographer's work from those particular magazines. The photographer should remember that a print of his making is not his property once it is first copyrighted by someone else, *unless* he has sold only certain rights of it. It is nothing less than theft, to make a photographic copy of a published photograph and to offer it as original and unpublished. The photographer should never try to sell what is not his own work. But since not many have the urge to do so, undue emphasis on that point would be offensive.

"The sum of the foregoing advice is that the author (photographer) should exercise common sense in disposing of rights," says J. Berg Esenwein, editor of the *Writer's Monthly*, in one of his books. "In most cases it would be better to allow the publisher to have 'All Rights' than to forego the chance of a sale; but nearly all magazine-editors are disposed to be reasonable and will agree to share any future profits that may arise from supplementary sales of a manuscript (photograph). The chief point is that author and publisher should clearly understand each other, without the author's losing his rights, yet, without harassing the publisher by making unnecessary stipulations regarding a trifling matter."

The law of copyright should be followed strictly when attempting to submit the same photograph to more than one publication or buyer. If the photographer keeps an eye on what rights he has sold when he cashes his cheque, and governs himself accordingly, he will sail along without trouble of any kind.

ILLUSTRATED SPECIAL ARTICLES

It would require a surveyor of extraordinary skill to mark the boundary between the lands of *Photographs-With-Explanatory-Data* and *Articles-Illustrated-With-Photographs*. Since the dividing line is so vague it is not difficult to pass from the one to the other.

The jump from the making of photographs to the writing of non-fiction is not a difficult one to make. In his rambles after salable photographs the press-photographer may unearth a subject to which a single photograph does not do justice. Then the making of more photographs and the writing of an article about them is the logical and the progressive and the more remunerative thing to do.

Indeed, subjects which would not sell otherwise may be made very useful to an editor by the writing of an enticing article around them. At once, there is a means of broadening one's market and of disposing of photographs, by themselves, unsalable. An illustrated article naturally calls forth a fatter cheque than would the text or the photographs alone. There is as much a demand for illustrated articles as there is for photographs; so that the photographer with the ability to tell facts simply and clearly has two avenues of revenue.

Many illustrated articles sold to magazines are just groups of photographs with interesting texts written about them. A search through a few magazines reveals a broad variety.

From *Popular Mechanics*:

- New Mountain-Road Now Open to Traffic.

- New Orleans Public Elevator.

- Artistic Roof-Garden Features City-Factory.

- Steamer Repaired in Eighteen Days.

- Where the Earth Collapsed.

- Flying Anglers Troll for Deep-Sea Fish.

- A Four-Track Concrete Railroad-Bridge.

- Waterfalls Near Big City Just Discovered.

- Concrete Smokestack Difficult to Demolish.

- Vast Stores of Mineral Paint-Pigments in Salton Sea.

From *Illustrated World*:

- What the Circus Does in Winter.
- Snow on the Overland Trail.
- City over Coal-Mines Slowly Sinking.
- Running the Farm by Windmill.
- Truck Equipped for Sealer of Weights and Measures.
- Marvelous Development in the Hemp-Industry.
- Public Camp-Conveniences.
- Mud-Splashing Guards for Autos.
- Work for Waterfalls Everywhere.
- Building the Road to Fit the Car.
- Heading Off Mountain-Floods.
- Lawn-Pools and Fountains in Concrete.

From *Photo-Era Magazine*:

- Children in the Snow.
- The Quartz-Meniscus Lens.
- Introduction of Figures in Landscape-Work.
- Photographic Greeting Cards.
- Balance by Shadows in Pictorial Composition.
- Mounting and Framing Photographs.
- The Photographer and a Goat-Ranch.
- In Nature's Studio.

From *Science and Invention*:

- Science Measures the Athlete.
- World's Largest Clock.
- Making Microphotographs.
- How Cartoon Movies are Made.
- A Miniature "Sky."
- Curing Soldiers' Ills with Electricity.

- Largest Electric Crane Lifts Complete Tug-Boat.

- Wintertime Uses for the Electric Fan.

- Monster Italian Searchlight.

These are articles written around several photographs—not merely illustrated by them. Besides the classes of magazines mentioned there are numerous others—almost any publication that uses illustrations in fact—which are in the market for illustrated articles. Such magazines cater to outers, hunters, sportsmen, business-men, physical culturists, travelers—almost every class of reader.

Having produced and sold articles written around the illustrations, the writer-photographer cannot other than form an idea, now and then, of an article a magazine should want which may be illustrated; but to which the illustrations are supplementary rather than basic. In such cases, the writer will have greater chance of acceptance if he, by means of his camera, makes several photographs to illustrate the text.

Even if an article is acceptable without illustrations, it will bring a bigger cheque nevertheless if it is illustrated. If the lack of illustrations makes the article unavailable, then the photographer has the means of making a cheque grow where none grew before. His camera stands him in good stead. There is no editor but prefers an illustrated article to an unillustrated one—unless his magazine is pictureless from policy.

Then, from having his pictures printed without his name attached, the photographer blossoms into a writer whose work appears under such a head as "'How Fruit is Raised on the Moon,' by John Henry Jones, with Illustrations by the Author."

Although the jump from the making of photographs to the writing of non-fiction is easy, you may slip at the first attempt. But hammer away and soon the nail will go in. "For know ye, there isn't a magazine-editor in the business who wouldn't buy an article from his worst enemy if he thought it was good stuff for his magazine."

The photographer must not only "smell out" news; but he must, by the sensitiveness of his "nose" tell just how much the news is capable of being worked up. He will find it comparatively easy to write illustrated special-articles where before he sold just photographs. And such ability stands not far below that of the fictionists.

THE HIGH ROAD

Not much of an exalted vocation, the selling of photographs? Not, perhaps, proclaimed from the housetops as a handsomely paying vocation; but one which may be cultivated into almost anything having to do with inveigling publishers into writing cheques.

When you receive your first cheque your sensation is something like that of the man who has passed through a cyclone and has come through with his "flivver" still in the barn. But when the first contribution is *printed!* The world is yours! You have broken into print! If not into type, at least into printing-ink.

When the excitement wears off there are many branches that beckon. The press-photographer may specialise—he may devote all his efforts to some one branch of the work, as the making of photographs of celebrities, of microphotographs, of almost anything. Witness the amateur photographer who quietly went about photographing the interior of every church in New York, and who then "cashed in" on them to the amount of $4,000. You may even obtain a position—or job—as press-photographer on a big metropolitan daily, with all the world before you and part of it dropping every Saturday afternoon into your pocketbook.

Then, you may be sent overseas—and be paid great oodles of money. Or you may devote all your time to the making of calendar-photographs, or to illustrating stories photographically, as is the fashion now with some magazines, see *True-Story*. There are so many opportunities to grasp that if you look about you and select the specialised branch in which you desire most to work, there is no reason in the world why you should not do it—and, perhaps, earn $10,000 a year at it. "Do one thing better than anyone else and the world will beat a path to your door."

Having broken into printers'-ink, it is comparatively easy to break into type. From selling photographs one may easily advance to the writing and illustrating of non-fiction. And your fame as a non-fictionist, together with the training you have gleaned, may cause you to forward a work of fiction to an editor acquainted with your name—and lo! from the ranks of the "snap-shooters" you have risen to the highest class of scribe—the successful fictionist.

And that, too, is not difficult for him who wills and works. "And work. Spell it in capital letters, WORK," advised Jack London. "Work all the time. Find out about this earth, this universe; this force and matter, and the spirit that glimmers up through force and matter from the maggot to Godhead. And by all this I mean work for a philosophy of life. It does not hurt how wrong

your philosophy of life may be, so long as you have one and have it well....
With it you may cleave to greatness and sit among the giants."

Another agrees: "Draw long breaths of confidence, of faith in yourself and
your work.... Strike 'despair' out of your dictionary! Get into your chair! Do
your stint! Be just as much of a fool as you like. It is your privilege and mine.
Then you will have amusing reminiscences. No great writer but can look back
and say, 'What a fool I was!'"

Realisation results from "ten per cent. inspiration and ninety per cent.
perspiration." A liberal quantity of this mixture will bring one to the High
Road. The High Road is smooth. But anyone may travel it who wishes—and
works sufficiently hard. Not much, the making and selling of photographs?
The start of the trail may be barren and unpromising; but the persevering
fellow who follows it persistently will find that it suddenly widens and
blossoms and lo, opens full into the High Road.

THE END